For Josephine

I have been thinking about the future.

As a climate scientist it is part of my job to explore what it might bring.

My career has been unusually varied.

It has included developing rocket and satellite instruments, initially to study the universe and the Sun, but then to observe the Earth, particularly its polar regions.

It has also included running the International Geosphere-Biosphere Programme that coordinates the activities of over 10,000 scientists in seventy-five nations.

I have been head of the British Antarctic Survey, and in that role have been to the Antarctic and the Southern Ocean many times.

I was President of the international scientific body that coordinates research in the Antarctic, and was one of the architects of the International Polar Year 2007–2008 – an exercise by thousands of scientists from over sixty countries to build up a 'snapshot' of the polar regions.

My work has enabled me to travel to parts of the planet visited by only a few and to meet experts from all over the world. This has allowed me to see and assess things for myself.

As Director of the Science Museum I moved away from running research projects and instead sought out ways for scientists and the public to discuss complex and controversial subjects.

Perhaps the most complex and controversial subject of all is that of our climate: if it is changing, in what way and on what timescale.

It is an extremely emotive issue, and we are all susceptible to bias and irrationality when confronting it.

The issues are often oversimplified, yet it is a subject of enormous complexity.

The climate system itself is very complicated, the most complex system we know of.

There are gaps in our knowledge, and many scientific uncertainties, some of which are fundamentally unknowable.

This makes it extremely difficult to predict precisely what the future holds and to determine exactly what, if anything, we should do.

In addition there are economic considerations, political implications and ethical questions that are not easily answered.

But decisions are being made on our behalf at various levels of government and **we all need to be part of that process**.

One of my current responsibilities, as Chair of the London Climate Change Partnership, is to draw together organisations within the public, private and civil sectors to make our capital city the best prepared and most resilient in the world with respect to climate change.

On some issues, such as flood protection, our planning extends one hundred years.

The future is also being prepared for at national and international levels.

In December this year, 195 nations will meet in Paris to agree on a course of action to respond to climate change.

Their discussions will be informed by a detailed summary of the latest climate science.

The decisions they make will affect us all.

I want to explain the results of the science, their implications, and the options we have before us.

A lot has changed in my lifetime.

When I was ten years old my mother gave me an atlas in which large areas of the Antarctic were marked 'Region Unknown to Man'.

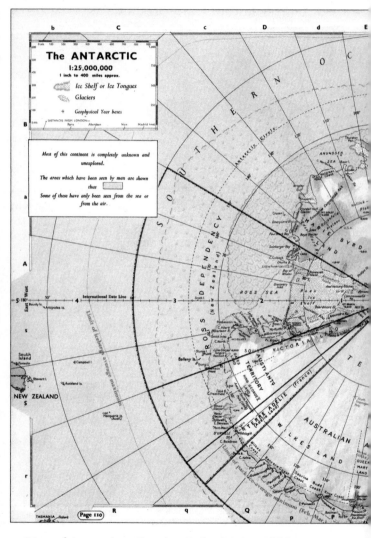

Map of Antarctica, *Concise Oxford Atlas*, 1958

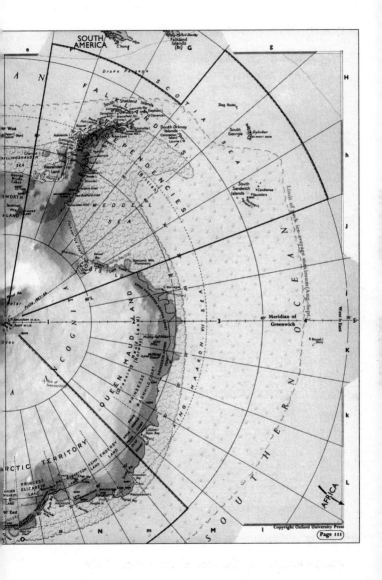

But in that same year an unprecedented sequence of scientific and technological advances began to open up that continent and render my atlas obsolete.

It was 1957, and the Commonwealth Trans-Antarctic Expedition embarked on the first ever crossing from coast to coast via the South Pole.

At the same time, despite the ongoing Cold War, sixty-seven countries, including the USSR and Eastern Bloc nations, collaborated in the International Geophysical Year – an intensive scientific campaign to study the Earth.

Major advances were made in oceanography, meteorology, magnetism and a host of other research fields.

One of the greatest advances was in the Antarctic where airborne surveys using ice-penetrating radars revealed, for the first time, the staggering depth of the ice sheet – up to 4 kilometres deep – and began to map out the mountainous terrain that lay below.

Then, on 4 October 1957, the Russians launched Sputnik 1, the first ever satellite.

I helped my father set up a short-wave radio set so that we could listen to Sputnik's faint bleeps among the hiss and crackle of static as it passed overhead.

Only four years after Sputnik, Yuri Gagarin became the first human in space, and just twelve years later – on 20 July 1969 – Neil Armstrong and Buzz Aldrin stood on the Moon.

By that time I had trained as a physicist at Oxford and was at home with my parents awaiting my degree result.

We watched the blurry footage of the Moon landing together on a Murphy black and white TV set.

A few years later, in 1971, I began my research career designing and building my own rocket and satellite instruments to study the cosmos.

I went on to work with NASA in America, designing and operating a satellite mission to study solar flares – explosive energy releases that occur in the Sun's atmosphere.

After six years, as we started work on a follow-up project, I saw some data from another pioneering NASA satellite mission called Seasat.

It was a contour map of the very edge of the Antarctic ice.

Instead of looking away from the Earth, Seasat looked back at it, using the radar instruments it carried to image and profile the oceans and polar regions.

It had the ability to map the shape of the ice sheet and, over time, to monitor its changes.

I knew I had to be part of this.

For the next fifteen years I built up a
research group specialising in the design
and use of radar altimeters to map the polar
regions from satellites.

At the same time I was a member of a small group of scientists from across Europe working with the European Space Agency to develop its series of Earth observation satellites.

This work culminated in the Cryosat satellite, which is operating to this day, taking hundreds of millions of measurements of the polar ice with pinpoint accuracy and unprecedented resolution.

Maps of Antarctica produced from Cryosat cover 96 per cent of the continent.

Very little of the region remains 'Unknown to Man'.

Height map of the Antarctic ice sheet from
the European Space Agency's Cryosat

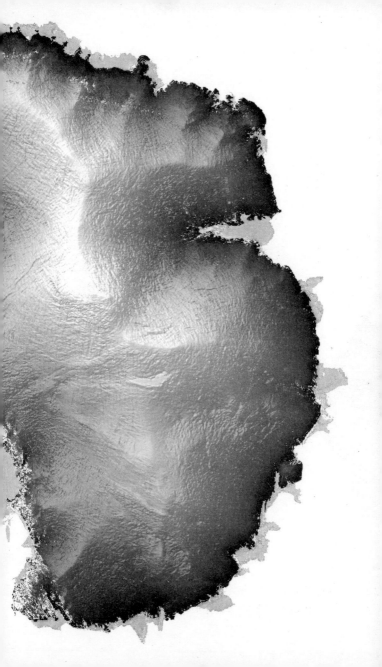

Polar research has not been alone in benefiting from satellite technology.

The advanced instruments available today allow us to probe and map the key components of the Earth – the atmosphere, the oceans, the ice and the land – in unprecedented ways.

For example, space radars are unaffected by cloud cover and darkness and, unlike human researchers on the surface, can continue to observe in the depths of the polar winter.

Imaging systems can resolve surface features as small as a metre across. And instruments with exquisite sensitivity can detect tiny changes in the Earth's gravity that allow us to measure changes in mass of the ice sheets and oceans.

We combine the space data with the myriad of measurements made from aircraft, ships, buoys and a host of specialised instruments on the ground.

And by using computer models to bring together the data with our understanding of the underlying physical laws, we can begin to make sense of what we observe.

This provides us with a grand perspective of the Earth's system as a whole, of its component parts and the interconnections between them.

The component parts are:

The atmosphere – the layers of gas
surrounding the planet

The hydrosphere – the oceans, lakes and rivers

The cryosphere – the ice on land and sea, the snow and the permafrost

The lithosphere – the outer layer of the rock that makes up most of the planet's mass

The biosphere – all living material,
including us

The system behaves in complex and often counterintuitive ways, but the fundamental principles of it are quite simple:

Its component parts interact with each other, exchanging energy in ways that operate in an overall **dynamic balance**.

Dynamic balance applies to many features of the system, such as the balance of carbon between the atmosphere, ocean, land and vegetation, or the amount of ice on land and on the water.

But it especially applies to the energy balance of the planet – meaning that, over time, the amount of energy leaving the planet is equal to the amount entering it.

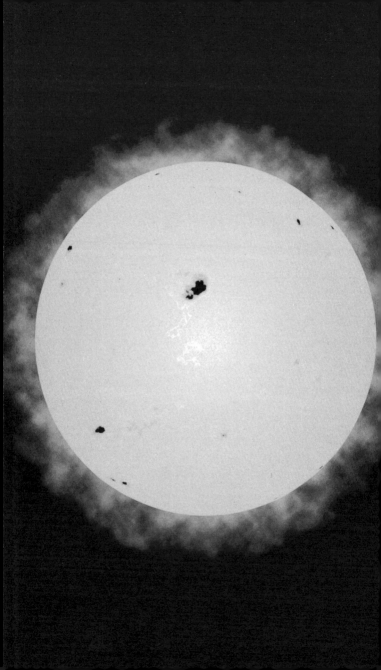

The primary source of energy is the Sun.

About a third of solar energy is reflected away by clouds, haze and the Earth's surface.

About a quarter is absorbed by the atmosphere.

Over 90 per cent of the remainder is absorbed by the oceans.

The rest of the energy goes into the land and ice.

Because the equatorial regions face the Sun directly, more energy is accumulated there than at the poles. The action of redistributing this excess then drives the circulation of the oceans and atmosphere.

Heat energy radiated by the Earth's surface to space is partially absorbed in the atmosphere by trace gases – water vapour, methane and carbon dioxide.

The interactions between the atmosphere, the oceans, and the ice on land and sea drive the natural variability of the climate.

The Earth system is very responsive, and even a small change in one component can trigger a chain of consequences in the other parts.

When such changes alter the energy balance, the effects are felt throughout the entire system.

Such changes include variations in the energy received from the Sun – either from fluctuations in the solar brightness or from small variations in the Earth's orbit and tilt that alter its distance and orientation relative to the Sun.

They include increases or decreases in the Earth's reflectivity due to variations in the cloud cover, or volcanic eruptions that inject haze into the upper atmosphere, or changes in the surface cover of snow or vegetation.

And they also include changes in the atmospheric concentrations of water vapour, methane and carbon dioxide that alter the planet's loss of energy to space.

Whenever one of these changes takes place, the climate system adjusts until a new energy balance is reached.

Some changes are amplified. An especially important effect occurs in the polar regions where, as highly reflective ice and snow melt, dark, heat-absorbing land or ocean is revealed beneath.

This increases the rate of melting and intensifies the warming.

Since the majority of the energy is absorbed by the ocean, any imbalance would be most observable in the hydrosphere.

To detect an energy imbalance in the oceans, we can analyse data from the worldwide system of ocean buoys – known as the Argo floats.

Over 3,500 of these have been deployed by thirty nations throughout the world's oceans since the millennium.

The instruments they carry record temperature, pressure and salinity down to a depth of 2,000 metres.

We can combine these measurements with contemporary and historic data from ships and other systems of buoys to allow us to estimate the ocean heat content and its variations.

Additionally, we can measure sea level, which rises as temperature increases, acting as a global thermometer.

A history of global sea level can be constructed from studies of beach structures worldwide, and from archaeological data, for example the location and height of Roman-era harbours, all of which hold a record of past sea levels.

There is also information from a network of tide gauges installed at harbours on coastlines around the world over a century ago, to provide data on local tides for seafarers and civil engineers.

The number of such installations has increased over the years to nearly 300 sites, creating the official 'Global Ocean Sea Level Network'.

More recently – over the last two decades – satellite radar altimeters have revolutionised sea-level measurements.

These provide almost complete ocean coverage, and they have the ability to detect changes in the global average level to within millimetres.

The combination of all these data shows that over the last several thousand years, global sea level was virtually static.

However, in the late nineteenth century it began to rise.

Over the twentieth century the rate of rise averaged 1.8 millimetres per year.

And over the last two decades, the rate has increased to 3.3 millimetres per year.

This may not sound much – but it indicates that the dynamic energy balance of the climate system has been disrupted.

Global average sea-level rise from tide gauges and satellite radar altimetry

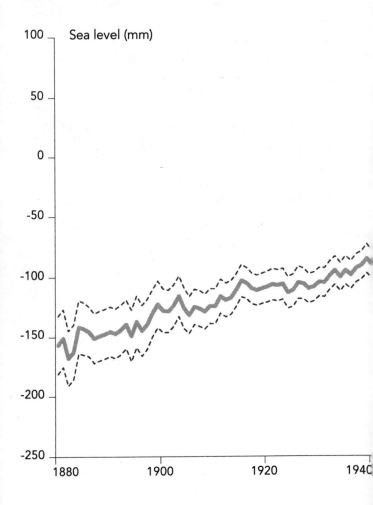

Sea level (mm)

Data from tide gauges, 1880 to 2009, yearly
Satellite altimeter data, 1993 to 2009, yearly

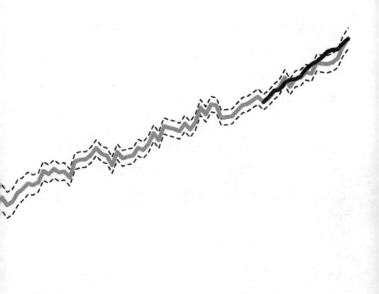

1960   1980   2000

To understand the implications of this imbalance, we have to put it in the context of geological time.

Stored in the rock and ocean sediments of the planet is a record of the past – often patchy and indistinct – but a record nonetheless.

When we investigate this record, we find that the world's climate has varied on many timescales and for many reasons.

Little is known of the early history of the planet after its formation 4.5 billion years ago.

We do know, however, that the biosphere emerged as soon as the physical conditions permitted – about 3.5 billion years ago.

Like the other components, the biosphere interacts with the rest of the system, in particular the atmosphere, in complex ways.

In the 'Great Oxidation', which started about 2.3 billion years ago, living organisms began producing oxygen in substantial quantities by photosynthesis.

This ultimately transformed the atmosphere into the oxygen-rich state that we experience today.

Over the last 500 million years the climate has varied between a warm state – much warmer than today – and a so-called 'icehouse state', in which ice sheets formed at one or both poles.

Between 360 and 300 million years ago, in the Carboniferous Period, conditions of temperature and moisture supported the formation of vast swamp forests.

In time their vegetation decayed, and was gradually overlain by sediments which, compressed and baked by geothermal heat, created deposits of coal, oil and gas.

Some 250 million years ago – in the Age of the Reptiles – the temperature of the planet was much higher than today, and so was the carbon dioxide concentration of the atmosphere.

The Age of the Mammals began about 65 million years ago following an asteroid impact in which more than three-quarters of all plant and animal species on Earth became extinct.

Life gradually recovered.

As it did so, the planet slowly cooled, and by about 34 million years ago ice sheets had developed at the South Pole.

The system then went through another warm phase, after which, about 20 million years ago, we entered the 'icehouse' world we experience today.

Over the last 2.5 million years the ice sheets at both poles have waxed and waned, initially on a 40,000-year cycle and, more recently – over the last million years – a 110,000-year cycle, triggered by small variations in the shape of the Earth's orbit and in its orientation to the Sun.

As the ice sheets wax and wane they have a huge impact on global sea level.

During the transition from the peak of the last ice age 18,000 years ago to the beginning of the current warm interglacial period, the oceans rose 120 metres at a sustained rate of about 1 metre per century – 10 millimetres per year.

Twelve thousand years ago, as the transition from the last ice age came to an end, the Holocene Epoch began.

Global average temperature stabilised, and since then has gently cooled by about 1°C, with fluctuations of no more than a degree.

The Holocene has been an extraordinarily stable period. We find nothing else like it in the climate record.

It is argued that this relative equilibrium has enabled our species to flourish, first establishing agriculture, then civilisation and then the modern world.

However, even the small climatic variations that have occurred during the Holocene have had significant human consequences.

For example, over the northern hemisphere, between the tenth and thirteenth centuries, regional warmings of up to 1°C took place.

This was the Medieval Warm Period, when the Vikings established substantial settlements on Greenland.

It was followed by a cooling from the sixteenth century to the nineteenth century, known as the Little Ice Age, when the Thames was repeatedly frozen and the Viking settlements in Greenland were abandoned.

These variations were small compared with the changes that took place over the 110,000-year ice age cycles.

During the last 2,000 years, sea level changes did not exceed 0.2 millimetres per year.

In this context, the 1.8 millimetres per year observed last century, and the 3.3 millimetres per year we observe today, are geologically significant.

The current rate is approaching that which occurs during the transition from an ice age to a warm interglacial, a major climatic shift.

And it is occurring during the warm interglacial, and at a time unrelated to the natural ice age cycle.

To understand what is causing this change in the hydrosphere we can first look to the cryosphere, in particular the ice shelves.

Ice shelves are floating platforms of ice hundreds of metres, or sometimes even a kilometre, thick that extrude off the main ice sheet on the land.

Both the ice sheet and the ice shelf move towards the ocean under the action of gravity.

Ice shelves commonly run aground in the shallow continental waters, and when they do so their movement is slowed, causing a build-up of pressure in the ice sheet behind them.

If ocean temperatures rise, the water warms the underside of the ice shelf, which then collapses into smaller sections – icebergs.

As the ice shelf collapses, the pressure on the ice sheet behind is relieved causing the ice sheet to flow more rapidly into the ocean, raising the sea level as it does so.

In 1995 the most northerly ice shelf of the Antarctic Peninsula collapsed.

For the first time in thousands of years it was possible to sail a ship around James Ross Island, which until then had been linked to the Trinity Coast of the Peninsula by the ice shelf.

The northern Peninsula ice shelves have waxed and waned even during the Holocene, so this event was not necessarily climatically significant.

However, since then, there has been a progressive southerly wave of collapses: the Larsen A ice shelf in 1995, the Wilkins in 1998, the Larsen B in 2002.

The Larsen B collapse was particularly significant, since evidence from sediment cores made accessible by its loss shows that it had been in place since the last Glacial Maximum 20,000 years ago.

In 2008 and again in 2009 parts of the Wilkins ice shelf – the largest on the Peninsula's west coast – collapsed. In two years it reduced to one-third of its original size.

I flew over the Wilkins ice shelf in 2009 and looked down on the vast area of shattered ice – it looked like pieces of a broken window.

The pilot, who had been flying in the region for more than twenty years, said he'd never seen anything like it.

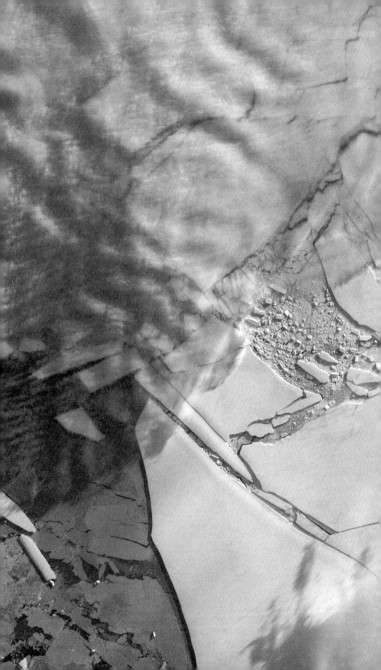

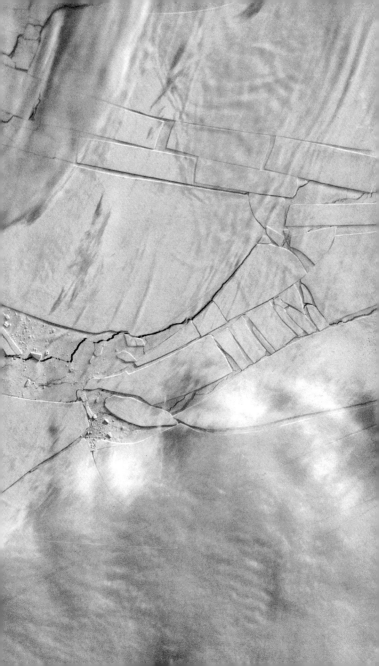

In 1978 John Mercer, a US glaciologist with much Antarctic field experience, described how, in a warming world, we might see a successive collapse of ice shelves extending down the Antarctic Peninsula.

He suggested that this would be a warning sign of a more worrying sequence of events to come.

Previous page: The Wilkins ice shelf on the western side of the Antarctic Peninsula

The Antarctic Peninsula connects to an area of the Antarctic called West Antarctica.

And the chief feature of West Antarctica is that its massive ice sheet sits on bedrock that is up to 2 kilometres below sea level.

For this reason it is called a marine ice sheet.

Mercer's concern was that if the successive collapse reached as far as West Antarctica, the pressure of the warmer water at depth would lift the ice sheet, causing water to penetrate deeper and deeper below the ice, reducing friction between the ice and rock, and so leading to an unstoppable collapse.

This would result in a rise in sea level over time of many metres, since the total volume of ice in West Antarctica is equivalent to a 6-metre rise.

As Mercer feared, the ice shelf collapses along the Peninsula have occurred and parts of the West Antarctic ice sheet are now starting to collapse.

Disturbingly, in East Antarctica, the Totten glacier, which is also marine-based, is showing accelerated ice loss too, which no one predicted.

And in the northern hemisphere, the satellite and surface data show that the loss of ice from the Greenland ice sheet increased by 600 per cent from about 34 gigatons per year in the late 1990s to 215 gigatons per year just a decade later.

One gigaton is one thousand million tons.

Greenland's fastest-flowing ice outlet, the Jakobshavn glacier, is now flowing in the summer at speeds of 17 kilometres per year – nearly 50 metres per day – the swiftest rate for any polar glacier or ice stream that has been recorded.

This glacier drains about 7 per cent of the ice sheet, and about 35 billion tons of icebergs calve off and pass out to sea every year.

It was one of these that is thought to have sunk the *Titanic* in 1912.

In addition to this ice loss in the Antarctic and Greenland, satellite image data reveal that 90 per cent of the world's glaciers and small ice caps are shrinking.

There is evidence from historical records that some glaciers, especially in Europe, began retreating as long ago as the mid-nineteenth century, probably in response to the end of the Little Ice Age.

But the satellite data show that the retreat globally has gathered pace over the last thirty years.

By analysing variations in the Earth's gravity field due to changes in the mass of the ice sheets and the ocean, it has been calculated that, at present, the melting of ice sheets and glaciers contributes about half of the observed sea level rise.

Apart from a small contribution from human use of aquifers, the rest of the sea level rise is due to thermal expansion.

In other words, the water is getting warmer.

We know that this warming is not due to the Sun's brightness increasing, because satellite instruments have been measuring the solar energy flux very accurately since the late 1970s.

We also find that while the lower atmosphere and surface are warming, the upper atmosphere is cooling.

If the Sun were the cause, the upper atmosphere would be warming too.

To understand what is causing the ice to melt and the oceans to warm, we need to look at what is happening to the atmosphere, in particular the trace gases – water vapour, methane and carbon dioxide.

These gases are present in relatively small quantities in our atmosphere in comparison to nitrogen and oxygen but they have a significant impact on the temperature of the planet.

Water vapour, methane and carbon dioxide obstruct the loss of heat from the surface as it passes upwards.

This effect, referred to as the 'greenhouse effect', causes the Earth's surface to have an average temperature of 15°C.

Without it, the surface would be -15°C.

Life as we know it would be impossible.

We can observe the change in atmospheric concentration over time by looking at data from ice cores drilled from ice sheets and glaciers in the Antarctic and Greenland.

The ability to study these ice cores is regarded by many as the most important advance in Earth science of the twentieth century.

Each year the snowfall creates a layer that compacts to ice and traps bubbles of the contemporary air.

The deepest ice cores extracted from the Antarctic are more than 3 kilometres long, and contain a record stretching back 800,000 years.

As the Director of the British Antarctic Survey, on one of my Antarctic trips in 2002 I visited the European drill site at a place called Dome C.

I watched as a 5-metre section of ice core, nearly half a million years old, was extracted from a depth of just under 3 kilometres.

It took an hour to lower the drill, a few minutes to drill the core section, and an hour to winch it up to the surface.

There are offcuts, small chunks of ice debris brought up with the core, which aren't useful to science.

I picked up a piece.

As a scientist I try to remain objective and dispassionate.

But here I was, in a part of the world that had fascinated me as a child when I looked at that area identified as 'Region Unknown to Man', holding a piece of ice that had not seen the light of day since before the dawn of mankind.

I listened to the air bubbles pop and crackle as the ice melted from the heat of my hand.

I breathed the air coming out of it, air that was trapped at the time of freezing.

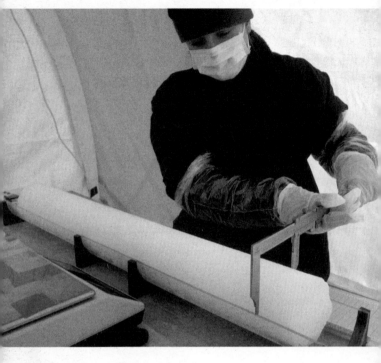

Processing a section of
Antarctic ice core

Bubbles of ancient air
trapped in a slice
of ice core

By measuring this air it is possible to study the composition of methane and carbon dioxide over time.

We melt the ice and measure the ratios of the different isotopes of hydrogen and oxygen in the water, which provide us with a history of global temperature.

We can then study the relationship between the trace gases and temperature.

The ice core data show an almost perfect match between the time curves of global temperature and atmospheric concentrations of carbon dioxide and methane.

As temperature increases, carbon dioxide and methane are released from the ocean and land biosphere, causing further temperature rise through their enhancement of the greenhouse effect.

The opposite takes place during the cooling phase.

Timeline of the variations over the last eight glacial cycles of atmospheric carbon dioxide (upper curve) and temperature (lower curve) derived from an Antarctic ice core

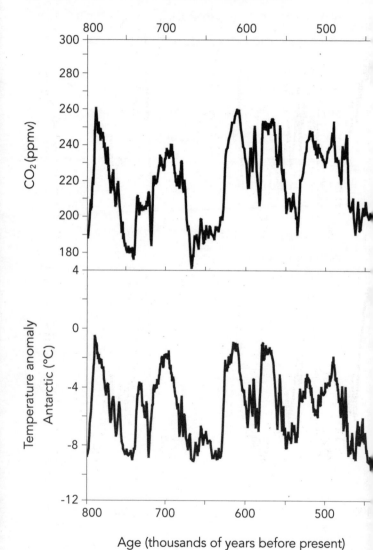

Age (thousands of years before present)

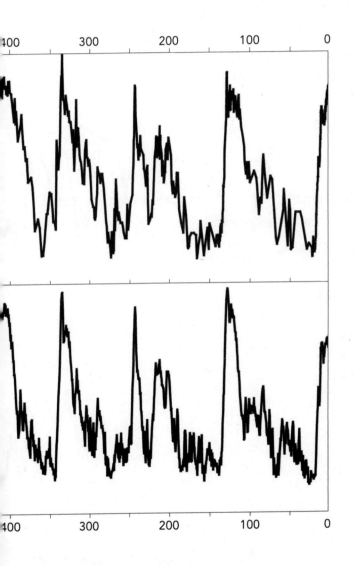

During the cold phase of each recent glacial cycle, when on average global temperatures have decreased by 5°C, the carbon dioxide concentration of the atmosphere dipped to about 180 parts per million.

In the warm phases it peaked at around 300 parts per million.

Last year, the carbon dioxide concentration of the atmosphere passed 400 parts per million.

Take a deep breath.

We're the first human beings to breathe air with that level of carbon dioxide.

It is unprecedented in the recent record.

The rise over the last century is already 100 parts per million – the same as the natural change between an ice age and an interglacial warm period, but at a rate more than a hundred times faster.

And it is in the 'warm' direction of increased concentration not experienced by the planet with certainty over the last 800,000 years based on the ice core data, and probably over 2 million years from the geological record.

The atmosphere is warming because the global carbon cycle has been disrupted.

The global carbon cycle consists of large annual exchanges between the carbon reservoirs of the atmosphere, the land biosphere, the lithosphere and the ocean.

These exchanges occur as a result of a variety of chemical, physical, geological and biological processes.

For example, as plants grow on land and in the sea in the spring, they draw down carbon dioxide from the atmosphere, which is later released as the green matter dies and decays.

Within the ocean, biological processes cause a fine rain of carbon to descend to the sediments, where it becomes trapped and stored.

Volcanic eruptions ultimately return lithospheric carbon dioxide to the atmosphere.

Physical exchanges take place between the atmosphere and the ocean as carbon dioxide is absorbed into cold dense waters that sink to depths, and is released from areas where warmer water upwells.

These exchanges are much greater in magnitude than our own carbon emissions – but prior to industrialisation they were in dynamic balance.

In 1712 the invention of the Newcomen steam engine started a chain reaction of innovation, technology and science that spread across the globe, driven by a desire for profit and the pursuit of a better life.

This revolution built the modern world.

It has been fuelled by cheap and accessible fossil energy, which had been accumulated over hundreds of millions of years during the Carboniferous Period, and stored underground as coal, oil and gas.

Since the 1950s, population, GDP, fertiliser use, water use, the number of cars, airline travel and many other human activities have all increased in what has been called 'The Great Acceleration'.

This has led us to the point where we are currently burning 10,000 million tons of carbon per year – a figure that has been increasing at a rate of 2 per cent per year.

To date, we have burned an estimated 530,000 million tons of carbon.

A quarter of the resulting carbon dioxide has been absorbed by vegetation on land, which has flourished as a result, and just over a quarter by the ocean, which has become more acidic.

The remainder will stay in the atmosphere for hundreds to thousands of years because it takes that long for natural processes – mainly rock weathering – to draw carbon dioxide out of the atmosphere.

Consequently, since the beginning of the Industrial Revolution, atmospheric concentration of carbon dioxide has risen by 40 per cent.

Human impact on the planetary system has been so profound that many feel we have irreversibly brought the climatic stability of the Holocene to an end and entered a new epoch.

The 'Anthropocene'.

The energy imbalance revealed by the ocean, confirmed by rising temperatures and loss of ice, is being driven by us. It is the unwitting consequence of our use of fossil fuels.

To me, the evidence seems compelling.

But the implications are so profound
that a more detailed and comprehensive
evaluation is merited.

To make such an evaluation requires a
gargantuan effort.

This is the task given to the Intergovernmental Panel on Climate Change – the IPCC – which was set up in 1988 by the United Nations Environment Programme and the World Meteorological Organisation.

Its job is to provide a comprehensive summary of the scientific data to inform the policy decisions of the United Nations Framework Convention on Climate Change.

This is an international treaty negotiated by 195 nations at the Earth Summit in Rio in 1992.

Its objective is to 'stabilise greenhouse gas concentrations in the atmosphere at a level that will prevent dangerous human interference with the climate system'.

The IPCC has three working groups. Working Group I reviews and assesses the physical science information relevant to human-induced climate change.

Working Group II addresses the related impacts on people and the environment.

And Working Group III focuses on the policy options for adaptation to and mitigation of human-induced climate change.

Since its establishment the IPCC has produced five Assessment Reports, approximately one every five years.

Each consists of a lengthy technical report, and a brief Summary for Policy Makers, which is scrutinised and agreed by representatives of the governments participating in the IPCC process.

The most recent Working Group I report – the fifth – was released in September 2013.

It is arguably the most audited scientific document – and possibly the most audited document – in history.

The work was led by 209 scientists, who are regarded as the world experts in their respective fields.

They were supported by more than 600 'contributing authors' from thirty-two countries, and fifty review editors from thirty-nine countries.

Of the tens of thousands of publications sifted more than 9,200 were cited.

The authors responded to 54,677 comments from 1,089 reviewers worldwide. And the final text was approved by representatives from 195 governments.

The full Working Group I Technical Report has 1,535 pages and weighs 4.25 kilos.

So what do they conclude?

Concerning the atmosphere: each of the last three decades has been successively warmer at the Earth's surface than any preceding decade since 1850.

In the northern hemisphere the thirty-year period from 1983 to 2012 was likely the warmest in the last 1,400 years.

They add that the globally averaged combined land and ocean surface temperature measurements show a warming of 0.8°C over the period 1850 to 2012.

They note that despite the warming at and near the surface, the upper atmosphere has cooled, ruling out the Sun as the cause.

They note that an increase in the frequency of heatwaves and heavy precipitation events has occurred in many regions since the 1950s.

Concerning the cryosphere: the rates of loss of ice from the world's glaciers and from the Greenland and Antarctic ice sheets have all increased dramatically – especially over the last thirty years.

While glacier losses have increased globally by about 20 per cent, the ice sheet losses have increased by as much as 600 per cent.

They report that the summer minimum sea ice extent in the Arctic decreased over the last thirty years at a rate between 9 and 14 per cent per decade.

There is evidence that this level of ice retreat is unprecedented in the last 1,450 years.

In contrast, winter sea ice extent in the Antarctic has increased slightly, at a rate of about 1.5 per cent per decade.

This appears to be driven by changes in the southern ocean winds, which have intensified in response to the planetary energy imbalance.

Other evidence of warming is provided by the loss of northern hemisphere snow cover at a rate of nearly 12 per cent per year in summer, and increases in permafrost temperatures too.

The warming since the 1980s has been 3°C in parts of north Alaska, and up to 2°C in parts of Russia, where a considerable reduction in permafrost thickness and geographic extent has been observed.

Concerning the hydrosphere: ocean warming dominates the increase in energy stored in the climate system, accounting for more than 90 per cent of the energy accumulated between 1971 and 2010.

Over the period 1993 to 2010 global mean sea level rise is consistent with a 39 per cent contribution from thermal expansion, 48 per cent from melting ice and 13 per cent from a decrease of land-water storage.

They confirm that the ocean has absorbed about 30 per cent of the cumulative anthropogenic carbon dioxide emissions, causing it to become progressively more acidic.

In November last year the IPCC released its overall Synthesis Report.

The report states that: 'Warming of the climate system is unequivocal, and, since the 1950s, many of the observed changes are unprecedented over decades to millennia.

'The atmosphere and ocean have warmed, the amounts of snow and ice have diminished, and sea level has risen.'

It observes that: 'In recent decades, changes in the climate have caused impacts on natural and human systems on all continents and across the oceans.

'Impacts are due to observed climate change, irrespective of its cause, indicating the sensitivity of natural and human systems to the changing climate.'

On the causes of climate change, the IPCC states that solar influences cannot account for the observations.

It adds that: 'It is extremely likely that more than half of the observed increase in global average surface temperature from 1951 [to] 2010 was caused by the anthropogenic increase in greenhouse gas concentrations and other anthropogenic forcings together.'

It continues that 'the best estimate of human-induced contributions to warming is similar to the actual warming observed.'

In other words, there is evidence that ALL the warming that has occurred since 1950 is due to human actions – due to us.

It concludes that: 'Continued emissions of greenhouse gases will cause further warming and changes in all components of the climate system.

'Limiting climate change will require substantial and sustained reductions of greenhouse gas emissions, which together with adaptation can limit climate change risks.'

John Kerry, the US Secretary of State, summarised the findings as follows: 'Boil down the IPCC report and here's what you find: Climate change is real, it's happening now, human beings are the cause of the transformation, and early action by human beings can save the world from its worst impacts.'

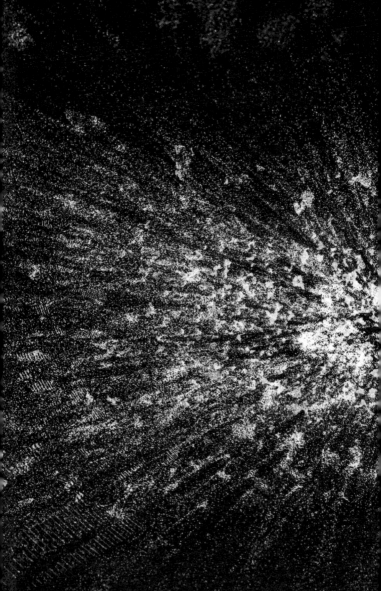

The cut-off date for published material considered by the IPCC Working Group I was July 2013.

But science never stops, and there have been some important results since.

Evidence from the Argo floats and ship-borne surveys shows that despite a fifteen-year pause in the rate of surface and atmospheric temperature rise, energy has continued to accumulate in the oceans unabated, with the prospect that some of this energy will be released to the atmosphere in the future.

New data from the Cryosat satellite show the recent rate of ice loss from Greenland and Antarctica has doubled in just three years.

Some experts have concluded that the loss of ice from the West Antarctic ice sheet is now irreversible, and that this will raise the sea level by 1–2 metres in as little as a few hundred years.

Based on a combination of scientific analysis, assessments of the impacts and related value judgements, the countries negotiating under the terms of the United Nations Framework Convention on Climate Change have set a limit beyond which climate change will be 'dangerous'.

That limit is 2°C above the pre-industrial average.

We are currently at 0.8°C.

Two-thirds of that increase has occurred since 1980.

In order to stay below the 2°C 'Guardrail', the science indicates that human carbon emissions have to drop to 50 per cent of the present level by 2050 and thereafter drop to zero.

Nothing.

This would mean leaving 75 per cent of known fossil fuel reserves in the ground. They would become economically worthless.

The temperature at which the system will stabilise is determined by the total quantity of carbon we emit to the atmosphere, not the rate at which it is emitted.

So reducing carbon emissions to zero won't lower temperature, it will just prevent the temperature rising beyond the 2°C level.

Temperature will then remain at that 2°C level for a very long time because carbon dioxide remains in the atmosphere for hundreds to thousands of years.

This sets a limit on the total carbon that we can burn.

The IPCC calculates this to be 800 gigatons of carbon.

It estimates that we have already burned 530 gigatons of carbon.

This leaves 270 gigatons for us to use.

At our current rate, which is 10 gigatons of carbon a year, we only have twenty-seven years left, after which time carbon emissions would need to cease.

Suppose we begin reducing our emissions this year and don't exceed the overall 800 gigaton limit:

Then the atmospheric carbon dioxide concentration will stabilise at 450 parts per million.

Temperature will take longer to stabilise because it responds to carbon dioxide concentration – but it will eventually stabilise at our Guardrail of 2°C.

CO₂ concentration, temperature, and sea level continue to rise long after emissions are reduced

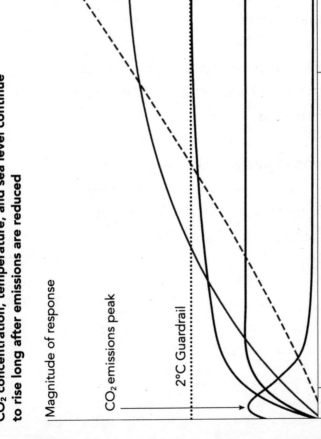

Magnitude of response

CO₂ emissions peak

2°C Guardrail

Today   100 years

1,000 years

Time taken to reach equilibrium

Sea-level rise due to ice melting: several millennia

Sea-level rise due to thermal expansion: centuries to millennia

Temperature stabilisation: a few centuries

CO₂ stabilisation: 100 to 300 years

CO₂ emissions

Despite this, the oceans will continue to warm and the ice will continue to melt.

So the sea level will continue to rise.

It will take hundreds of years but will eventually stabilise at a level, based on evidence from past warmings, some 2–3 metres higher than today.

If we leave it longer to start reducing our emissions, we'll have to reduce them more rapidly to avoid exceeding the overall 800 gigaton limit.

Calculations show that if we leave it until 2020 – only five years away – the subsequent reductions would have to be of the order of 6 per cent per year – year on year – to stay within the 2°C limit.

Six per cent may not sound much but annual reductions of carbon emissions greater than 1 per cent have historically been associated only with economic recession or upheaval.

The UK conversion from coal to gas and the French conversion to nuclear in the 1970s and 1980s achieved reductions of only about 1 per cent per year.

A temporary 5 per cent reduction was achieved in the Soviet Union when it collapsed.

The 6 per cent sustained rate of reduction required is of global emissions.

We in the developed world have to reduce our emissions even more rapidly to accommodate growth in the developing world.

In 2014 the UK achieved a 9 per cent reduction in emissions, as a result of a combination of a 20 per cent shift from coal to gas, an unusually warm year, and a fall in energy use.

This shows that progress on this scale can be possible within a national context.

But to achieve the necessary emissions trajectory will require a massive collaborative effort on a global scale.

It will require the greatest collective action in history.

In December 2015, 195 nations will meet in Paris to forge a deal to put the world on to a path to a 2°C maximum rise.

The new agreement aims to obtain credible and fair emission reductions and legally binding commitments from all countries – reflecting GDP, mitigation potential and contributions to past and future climate changes – with the most advanced economies making the most ambitious commitments.

There is justified cynicism surrounding the Paris meeting.

These nations have been meeting for decades and overall global emissions have not yet decreased.

However, there are hopeful signs from world leaders and governments and growing pressure on them from an increasingly informed populace.

Last year, a million people around the world marched in various capital cities to demonstrate their concern.

In the lead-up to Paris 2015, and prior to the recent talks in Lima, President Obama and Chinese President Xi Jinping announced joint measures to fight climate change.

The US aims to reduce its carbon emissions to 26–28 per cent below 2005 levels by 2025 – nearly doubling its previous commitments.

Despite not having ratified the Kyoto Treaty, the USA is already on track to cut its emissions by 17 per cent between 2005 and 2020.

China, partly driven by serious air pollution problems, has committed to cutting the proportion of energy it generates from coal and has set up pilot carbon markets and low carbon zones.

It has set a date of 2030 for when it plans to 'peak' its emissions, and has pledged to increase the share of non-fossil fuels in its energy mix to around 20 per cent by 2030, from less than 10 per cent today.

Prime Minister Narendra Modi of India has committed to expand solar energy to provide electricity to 300 million of his country's citizens, who have no access to power at present.

The European Union has agreed a package to achieve a 40 per cent reduction in its domestic emissions by 2030.

It aims to boost the use of renewable energy to 27 per cent and to increase energy efficiency by at least 27 per cent.

And Germany has committed to curbing its emissions by 40 per cent by the end of 2020, with the longer-term goal of supplying 80 per cent of its power from renewable sources by 2050.

The UK Climate Change Act, passed in 2008 with cross-party support, is the world's first long-term, legally binding, national framework for reducing emissions – setting five-year carbon budgets to cut UK emissions by 80 per cent by 2050.

Around the world, almost 500 climate-related laws have been passed in sixty-six of the world's largest emitting countries.

Although the long history of negotiations by the United Nations Framework Convention on Climate Change is regarded by many as a chapter of failures, others argue that it has created the conditions in which national legislators and decision-makers have been able to take actions that have had a positive effect.

In 2005 the mayors of the world's forty largest 'Megacities' – including London – met and formed the C40 Cities Climate Leadership Group.

These cities have a combined population of 297 million people, and they generate 18 per cent of global GDP as well as 10 per cent of global carbon emissions.

Collectively they have taken 4,734 actions to tackle climate change. Over three-quarters of these actions have been implemented.

Many individuals have taken measures to reduce their own climate-related impacts by making changes in their personal, professional and public lives: installing solar panels; increasing the energy efficiency of their homes, vehicles and appliances; using public transport and avoiding unnecessary travel; changing diet and choosing to forgo activities that generate emissions.

They have encouraged changes to be made in their workplaces and written to their MPs.

They have sought to educate themselves about the issue and to talk about it with their friends, families and communities.

But despite all these measures, global carbon emissions continue to rise. In 2013 they increased by 2.1 per cent to reach 10 gigatons of carbon per year, their highest value yet.

Suppose we fail to take the action needed to stay below the 2°C Guardrail?

The IPCC Working Group I predicts that by the end of the century we could have committed to a rise of more than 4°C.

A 4°C world would be one of unprecedented heatwaves, severe drought, and major floods in many regions, with serious impacts on ecosystems and food and water supplies.

Given that uncertainty remains about the full nature and scale of impacts, there is no certainty that adaptation to a 4°C world is possible.

A 4°C world is likely to be one in which communities, cities and countries would experience severe disruptions, damage and dislocation, with many of these risks spread unequally.

The International Energy Agency's 2012 assessment indicated that without further mitigating action there is a 40 per cent chance of warming exceeding 4°C by 2100 and a 10 per cent chance of exceeding 5°C.

No nation would be immune to the impacts of that level of climate change.

Our infrastructure was built for the climate system we inherited, and is not designed to cope with the climate system we are provoking.

Our food and water supplies, housing, industry – our entire wellbeing and prosperity – depend on access to energy, and our primary source, at present, is fossil fuel.

So we are confronted with a need to totally transform the world's energy system.

At the same time we need to ensure energy security, equity, sustainability and growth.

The amount of carbon we emit is determined by four things:

The number of people on the planet.

The size of the global economy.

The amount of energy it takes to power that economy.

And the amount of carbon it takes to create that energy.

History suggests that there's little we can do about population growth, which continues, albeit at a decelerating rate, and is projected by the UN to peak at about 9 billion later this century.

Similarly, there's little we can do to reduce the global economy. All governments are committed to increasing it and, in any case, our prosperity and wellbeing depend on it.

So there are only two areas in which we can take action to reduce emissions.

The economy can become more energy efficient and less wasteful.

This can be achieved through energy standards legislation, and by changes in behaviour at a personal and societal level.

Many relevant policies are already in place, and progress is being made.

But we haven't got close to the reduction we need.

Which leaves us with reducing the amount of carbon we emit as we generate energy.

This involves renewable power sources – wind, solar, biofuels, nuclear – and, if feasible, 'clean' – or 'carbon abated' – fossil fuels.

Around the world, renewable power capacity grew at its strongest ever pace in 2013 and now produces 22 per cent of world energy.

More than 250 billion dollars was invested in 'green' generating systems in 2013, although the growth is expected to slacken, partly because Western politicians are seeking to reduce financial incentives.

In contrast, in China authorities have set green energy as a strategic priority. Their aim is that it will account for more than half of China's energy production by 2050.

This explains why investors are increasingly confident and keen to put their money on alternative energy.

The growth rate of wind farms and solar plants in China, India and an array of smaller developing nations is starting to outpace that in the world's richest nations.

But, in the UK, despite our best efforts to move to green, renewable and nuclear, coal and gas still provide about 70 per cent of the energy supplied to our grid, and a higher percentage of that used in transport.

To achieve the necessary magnitude and rate of reduction in carbon emissions will require all the clean energy options available to us, as well as the invention and mass roll-out of new technologies, which, at this present moment, do not exist.

It is a daunting challenge, but my experience at the Science Museum, with its legacy of technical innovation on public display and held in its reserve collection and archives, convinces me that on a finite planet human ingenuity is unbounded.

My hope lies with the engineers.

But the right conditions need to be in place for innovation and exploitation to occur.

I would like to see governments, investors and the engineering profession itself create the conditions for a massive effort of innovation and roll-out of energy technologies that will make existing fossil fuel redundant – energy that is cheaper and cleaner than unabated fossil fuels.

Once it is available, the markets will drive its exploitation.

But progress is hard when other economic drivers inhibit the transformation.

Fossil fuels are estimated by the International Energy Agency to receive subsidies of 500 billion dollars per year – six times the incentives to develop renewables.

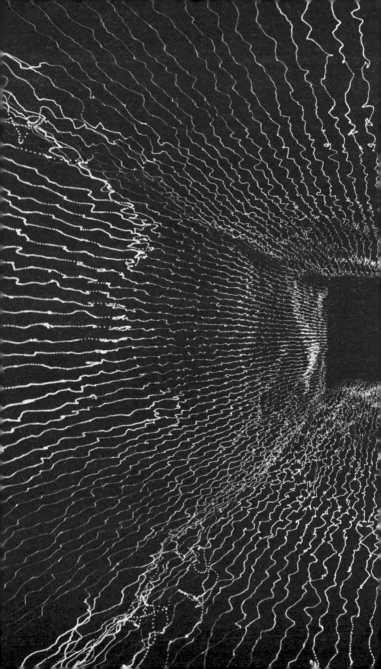

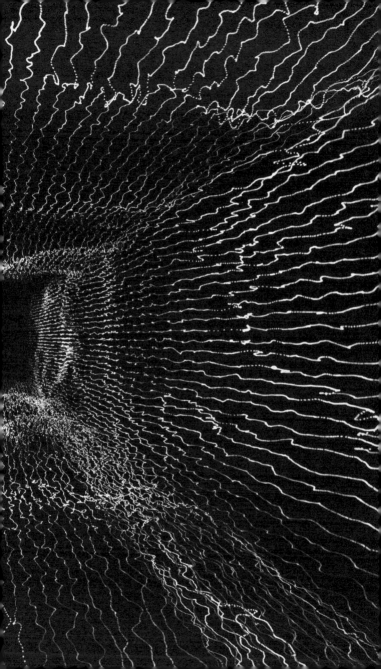

I think back to the remarkable collaborative efforts of the International Geophysical Year, culminating in the first satellite and the obsolescence of my childhood atlas – and the 'Region Unknown to Man' that I would go on to visit personally dozens of times.

I look at my eldest grandchild who is now the age I was during that world-changing year.

I tell her I think she should become an engineer.

She will reach the age I am now in 2071.

I try to imagine 2071, and then I find myself
thinking what 4071 will look like.

Or 10071.

We are all dependent on energy. Almost everything we do depends on it.

There will be carbon atoms generated by the production of this book that will still be in the air in 2071, in the air that my granddaughter will breathe.

That and all our other carbon dioxide emissions are our legacy.

Science cannot say what is right and what is wrong.

Science can inform, but it cannot arbitrate, it cannot decide.

Science can say that if we burn another half-trillion tons of carbon the carbon dioxide content of the atmosphere will increase by another 100 parts per million, and that will almost certainly lead to a warming of the planet greater than 2°C, resulting in major disruption of the climate system, and huge risks for the natural world and human wellbeing.

But science cannot answer moral questions, value questions.

Do we care about the world's poor? Do we care about future generations? Do we see the environment as part of the economy, or the economy as part of the environment?

The whole point about climate change is that, despite having been revealed by science, it is not really an issue about science, it is an issue about what sort of world we want to live in.

What kind of future do we want to create?

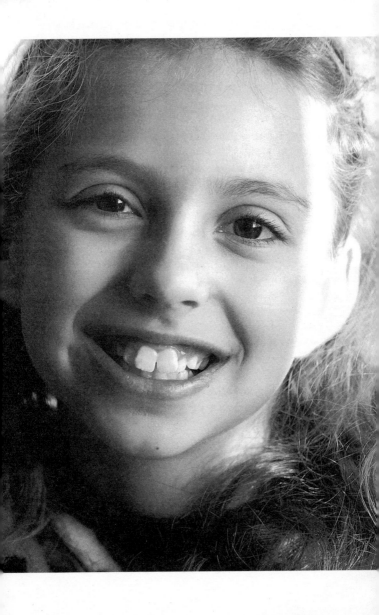

The authors wish to recognise the contribution of Katie Mitchell, who initiated the writing and production of the play, and whose energy and enthusiasm helped drive it to success. We also gratefully acknowledge the sponsorship and commitment of the Royal Court Theatre, London, and Deutsches Schauspielhaus, Hamburg.

Pages 16–17: Map of Antarctica from the Concise Oxford Atlas, 2E (1958) pp. 110–111 © Oxford University Press. Pages 28–9: Height map of the Antarctic ice sheet from Cryosat. Courtesy of Malcolm McMillan, CPOM, University of Leeds. Pages 34–38 and 42: © Shutterstock.com. Pages 58–9: Observed change in global mean sea level, 1880–2009. Data of global mean sea level since 1880 provided by CSIRO. Global and European sea-level rise (CLIM 012) assessment published 2014 © European Environment Agency (EEA). Pages 82–3: Cracking ice shelf. The Wilkins ice shelf, NASA image created by Jesse Allen, using data provided courtesy of NASA/GSFC/METI/ERSDAC/JAROS and U.S./Japan ASTER Science Team. Page 102: Processing a section of Antarctic ice core, European Project for Ice Coring in Antarctica (EPICA). © Chris Gilbert/British Antarctic Survey. Page 103: A close-up of a slice of ice core showing trapped air bubbles. © Pete Bucktrout/ British Antarctic Survey. Pages 106–7: Timeline showing variations of carbon dioxide and temperature derived from an Antarctic ice core, EPICA project. Courtesy of Eric Wolff. Page 155: Courtesy of IPCC. Climate Change 2001: Synthesis Report. A Contribution of Working Groups, I, II and III to the Third Assessment Report of the Intergovernmental Panel on Climate Change, Figure SPM-5. Cambridge University Press. Note addition of guardrail to original image. Page 205: Josephine, Chris Rapley's eldest grandchild. Photo courtesy of Chris Rapley. Graphics on pages 115, 144–5, 162–3 and 194–5 courtesy of Luke Halls, adapted from his graphics for the original stage production of 2071 at the Royal Court Theatre, 2014–15.

Chris Rapley is Professor of Climate Science at University College London. He is a Fellow of St Edmund's College, Cambridge, a Distinguished Visiting Scientist at NASA's Jet Propulsion Laboratory in Pasadena, California, a member of the Academia Europaea, a board member of the Winston Churchill Memorial Trust, Chairman of the European Space Agency Director General's High Level Science Policy Advisory Committee, and Chairman of the London Climate Change Partnership, committed to ensuring London's resilience to climate change.

His previous posts include Director of the Science Museum, London, Director of the British Antarctic Survey and Executive Director of the International Geosphere-Biosphere Programme at the Royal Swedish Academy of Sciences in Stockholm. Before that Professor Rapley established and built up the Earth Observation satellite group at UCL's Mullard Space Science Laboratory.

Professor Rapley was awarded the 2008 Edinburgh Science Medal for having made 'a significant contribution to the understanding and wellbeing of humanity'. He was made a Commander of the British Empire in 2003.

Duncan Macmillan is an award-winning writer and theatre director. His work has been selected for the Festival d'Avignon, Theatertreffen Berlin and the Salzburg Festival. His plays have been staged throughout the world including at the National Theatre, Royal Court, Almeida, Schaubühne Berlin and in London's West End.

First published in Great Britain in 2015 by John Murray (Publishers)
An Hachette UK Company

1

© Chris Rapley and Duncan Macmillan 2015

A CIP catalogue record for this title is available from the British Library

ISBN 978-1-47362-215-9
Ebook ISBN 978-1-47362-216-6

Book design and typesetting by Craig Burgess

Printed and bound by Clays Ltd, St Ives plc

John Murray policy is to use papers that are natural, renewable and
recyclable products and made from wood grown in sustainable
forests. The logging and manufacturing processes are expected to
conform to the environmental regulations of the country of origin.

John Murray (Publishers)
Carmelite House
50 Victoria Embankment
London EC4Y 0DZ

www.johnmurray.co.uk